2023 年

全国水利发展统计公报

2023 Statistic Bulletin
on China Water Activities

中华人民共和国水利部 编

Ministry of Water Resources of the People's Republic of China

·北京·

图书在版编目（CIP）数据

2023年全国水利发展统计公报 = 2023 Statistic Bulletin on China Water Activities / 中华人民共和国水利部编. -- 北京：中国水利水电出版社，2024.6
ISBN 978-7-5226-2481-5

Ⅰ．①2… Ⅱ．①中… Ⅲ．①水利建设－经济发展－中国－2023 Ⅳ．①F426.9

中国国家版本馆CIP数据核字(2024)第111326号

书　名	**2023年全国水利发展统计公报** **2023 Statistic Bulletin on China Water Activities** 2023 NIAN QUANGUO SHUILI FAZHAN TONGJI GONGBAO
作　者	中华人民共和国水利部　编 Ministry of Water Resources of the People's Republic of China
出版发行	中国水利水电出版社 （北京市海淀区玉渊潭南路1号D座　100038） 网址：www.waterpub.com.cn E-mail：sales@mwr.gov.cn 电话：（010）68545888（营销中心）
经　售	北京科水图书销售有限公司 电话：（010）68545874、63202643 全国各地新华书店和相关出版物销售网点
排　版	中国水利水电出版社微机排版中心
印　刷	北京印匠彩色印刷有限公司
规　格	210mm×297mm　16开本　3.25印张　56千字
版　次	2024年6月第1版　2024年6月第1次印刷
印　数	0001—1000册
定　价	**39.00元**

凡购买我社图书，如有缺页、倒页、脱页的，本社营销中心负责调换

版权所有·侵权必究

目 录

1 水利建设投资 ……………………………………………… 1

2 重点水利建设 ……………………………………………… 4

3 主要水利工程设施 ………………………………………… 7

4 水资源节约集约利用 ……………………………………… 10

5 防洪抗旱 …………………………………………………… 12

6 水利改革与管理 …………………………………………… 14

7 水利行业状况 ……………………………………………… 20

Contents

I. Investments in Water Resources Construction 25

II. Construction of Key Water Resources Projects 28

III. Key Projects and Structures 30

IV. Water Resources Conservation, Utilization and Protection 34

V. Flood Control and Drought Relief 36

VI. Water Reform and Management 37

VII. Current Status of the Water Sector 44

　　2023年是全面贯彻党的二十大精神的开局之年。一年来，各级水利部门坚持以习近平新时代中国特色社会主义思想为指导，深入贯彻党的二十大精神，认真践行习近平总书记"节水优先、空间均衡、系统治理、两手发力"治水思路和关于治水的重要论述精神，认真落实党中央、国务院决策部署，完整、准确、全面贯彻新发展理念，统筹高质量发展和高水平安全，坚持治水思路，坚持问题导向，坚持底线思维，坚持预防为主，坚持系统观念，坚持创新发展，水旱灾害防御夺取重大胜利，水利建设投资和规模创历史新高，水利改革创新取得突破性进展，各项水利工作圆满完成年度目标任务，推动新阶段水利高质量发展迈出坚实步伐。

1 水利建设投资

2023年，水利建设完成投资11996亿元。其中：建筑工程完成投资9089亿元，占75.8%；安装工程完成投资583亿元，占4.9%；机电设备及工器具购置完成投资321亿元，占2.7%；其他完成投资2003亿元，占16.6%。2016—2023年水利建设完成投资情况见表1。

表1　2016—2023年水利建设完成投资情况　　单位：亿元

按规模分	2016年	2017年	2018年	2019年	2020年	2021年	2022年	2023年
全年完成投资	6100	7132	6603	6712	8182	7576	10893	11996
建筑工程	4422	5070	4877	4988	6015	5851	8492	9089
安装工程	255	266	281	243	320	330	486	583
机电设备及工器具购置	173	212	214	221	250	204	287	321
其他	1250	1584	1231	1260	1597	1191	1628	2003

从项目类型看，流域防洪工程体系建设完成投资3227亿元，占完成总投资的27%；国家水网重大工程建设完成投资5665亿元，占完成总投资的47%；复苏河湖生态环境建设完成投资2079亿元，占

完成总投资的17%；水文基础设施、数字孪生水利建设等完成投资1025亿元，占完成总投资的9%。2023年分项目类型完成投资情况如图1所示。

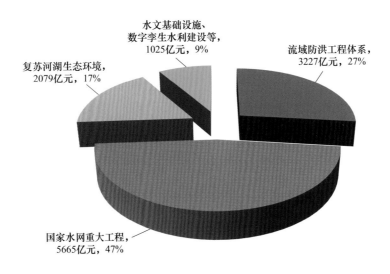

图1 2023年分项目类型完成投资情况

在全年完成投资中，中央项目完成投资118亿元，地方项目完成投资11878亿元。按项目规模分，大中型项目完成投资2987亿元，小型及其他项目完成投资9009亿元。按项目性质分，新建项目完成投资9167亿元，改扩建项目完成投资2829亿元。

全年实施水利建设项目 41014 个,项目投资总规模 54797 亿元。其中,新开工项目 27925 个,较上年增加 11.5%。2011—2023 年水利建设完成投资情况如图 2 所示。

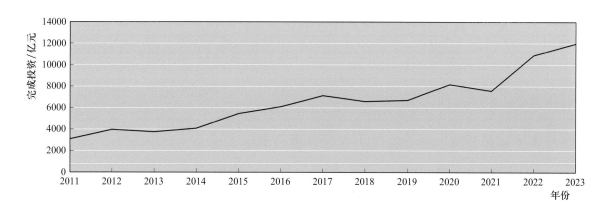

图 2　2011—2023 年水利建设完成投资情况

2 重点水利建设

流域防洪工程建设。长江流域姚家平、凤凰山等水库，长江安庆河段治理工程，鄱阳湖康山、珠湖等蓄滞洪区开工建设；黄河流域东庄水库、王瑶水库扩容工程加快建设，下游防洪工程建设全力推进；淮河流域重点平原洼地治理、沿淮行蓄洪区治理等开工建设，淮河入海水道二期工程加快实施；海河流域实施北京温潮减河，天津北运河，河北滹沱河、滏阳新河治理，青山水库开工建设；珠江流域梧州防洪堤、松花江流域十六道岗水库、太湖流域扩大杭嘉湖南排后续西部通道开工建设，吴淞江治理有序推进，太湖环湖大堤全线达标。

国家水网骨干工程建设。 吉林水网骨干工程、黑龙江粮食产能提升重大水利工程、雄安干渠、环北部湾广西水资源配置等44项重大水利工程开工建设。一批重大水利工程建设实现关键节点目标，西江大藤峡水利枢纽主体工程完工，引汉济渭先期通水，引江济淮试调水，广东珠三角水资源配置工程具备全线通水条件，引江补汉、滇中引水工程加快建设，黑河黄藏寺水利枢纽下闸蓄水。

农村水利基础设施建设。 云南腾冲、广西下六甲等5处大型灌区开工建设，598处大中型灌区续建配套与现代化改造加快推进，建成后将新增恢复和改善灌溉面积约7000万亩。开工建设农村供水工程2.3万处，提升1.1亿农村人口供水保障水平，农村自来水普及率达到90%，规模化供水工程覆盖农村人口比例达到60%。4000余座水电站完成了智能化改造，200多处小水电集控中心和分中心投运。新增农村水电站44座，新增装机50.65万千瓦。

重要河湖治理与修复。 构建丹江口库区及其上游流域水质安全保障工作体系。福建木兰溪、吉林查干湖、安徽巢湖等重点河湖水生态治理修复加快实施，断水百年之久的京杭大运河再度全线水流贯通，断流干涸26年之久的永定河首次实现全年全线有水。全面推进华北区域及其他重点区域地下水超采综合治理。全国新增水土流失综合治理面积7.0万平方公里，其中国家水土保持重点工程新增治理水土流失面积1.27万平方公里。综合整治坡耕地82.46万亩，新建淤地坝和拦沙坝600座，除险加固病险淤地坝432座。

数字孪生水利建设。数字孪生流域、数字孪生水网、数字孪生工程建设全面推进。全国首个数字孪生流域建设重大项目——长江流域全覆盖水监控系统开工建设。数据量达千万亿字节（PB）级的全国一级数据底板建设完成，部级水利模型平台、知识平台、资源共享平台建设取得明显进展。雨水情监测预报"三道防线"加快构建，流域防洪业务"四预"应用取得突破。国家水土保持监测站点优化布局工程立项建设。数字孪生灌区先行先试持续推进。

3 主要水利工程设施

水库和枢纽。全国已建成各类水库94877座,水库总库容9999亿立方米。其中:大型水库836座,总库容8077亿立方米;中型水库4230座,总库容1210亿立方米。

堤防和水闸。全国已建成5级及以上江河堤防32.5万公里,累计达标堤防25.7万公里,堤防达标率为79.0%;其中,1级、2级达标堤防长度为4.0万公里,达标率为88.3%。全国已建成江河堤防可保护人口6.4亿人,保护耕地41748×10³公顷。全国已建成流量为5立方米每秒及以上的水闸94460座,其中大型水闸911座。按水闸类型分,分洪闸7300座,排(退)水闸16857座,挡潮闸4522座,引水闸12461座,节制闸53320座。

机电井和泵站。全国已累计建成日取水大于等于20立方米的供水机电井或内径大于等于200毫米的灌溉机电井共521.6万眼。全国已建成各类装机流量1立方米每秒或装机功率50千瓦以上的泵站

91324处，其中：大型泵站507处，中型泵站5014处，小型泵站85803处。

灌区工程。全国已建成设计灌溉面积2000亩及以上的灌区共21340处，耕地灌溉面积40075×10³公顷。其中：50万亩及以上大型灌区154处，30万~50万亩大型灌区296处。截至2023年年底，全国灌溉面积80605×10³公顷，耕地灌溉面积71644×10³公顷，占全国耕地面积的56.1%。

农村水电站。截至2023年年底，全国现有农村水电站41114座，装机容量8157万千瓦，占全国水电装机容量的19.4%。全国农村水电年发电量2303亿千瓦时。

水土保持工程。全国水土流失综合治理面积达162.7万平方公里，累计封禁治理保有面积达32.5万平方公里。2023年持续开展全国全覆盖的水土流失动态监测工作，全面掌握县级以上行政区、重点区域、大江大河流域的水土流失动态变化情况。

水文站网。全国已建成各类水文测站127035处，包括国家基本水文站3312处、专用水文站5169处、水位站20633处、雨量站56279处、蒸发站9处、地下水站26576处、水质站9187处、墒情站5809处、实验站61处。其中，向县级以上水行政主管部门报送水文信息的各类水文测站84921处，可发布预报站2575处，可发布预警站2729处；配备在线测流系统的水文测站3537处，配备视频监控系统的水文测站7379处。基本建成中央、流域、省级和地市级共330个水质监测（分）中心和水质站（断面）组成的水质监测体系。

水利网信。截至2023年年底，全国省级以上水利部门累计配置各类服务器10495台（套），形成存储能力58.32拍字节，存储各类信息资源总量达6.37拍字节；县级以上水利部门累计配置各类卫星设备3766台（套），利用北斗卫星短文传输报汛站达10809个，应急通信车49辆，集群通信终端2827个，宽、窄带单通信系统464套，无人机3351架。全国省级以上水利部门各类信息采集点达60.35万处，其中：水文、水资源、水土保持等信息采集点约30.35万个，大中型水库安全监测采集点约30万个。

4 水资源节约集约利用

水资源状况。 2023年，全国水资源总量25782.5亿立方米，比多年平均值偏少6.6%。全国年平均降水量642.8毫米，比多年平均值偏少0.2%，较上年增加1.8%。全国768座大型水库和3942座中型水库年末蓄水总量4594.5亿立方米，比年初增加390.4亿立方米。

水资源开发。 2023年，新增规模以上水利工程❶供水能力91.8亿立方米。截至2023年年底，全国水利工程供水能力达9062.0亿立方米，其中：跨县级区域供水工程677.8亿立方米，水库工程2474.1亿立方米，河湖引水工程2119.7亿立方米，河湖泵站工程1852.6亿立方米，机电井工程1372.6亿立方米，塘坝窖池工程371.9亿立方米，非常规水资源利用工程193.3亿立方米。

水资源利用。 2023年，全年总供水量5906.5亿立方米，其中：地

---

❶ 规模以上水利工程包括：总库容大于等于10万立方米水库、装机流量大于等于1立方米每秒或装机容量大于等于50千瓦的河湖取水泵站、过闸流量大于等于1立方米每秒的河湖引水闸、井口井壁管内径大于等于200毫米的灌溉机电井和日供水量大于等于20立方米的机电井。

表水供水量 4874.7 亿立方米，地下水供水量 819.5 亿立方米，非常规水源供水量 212.3 亿立方米。全国总用水量 5906.5 亿立方米，其中：生活用水量 909.8 亿立方米，工业用水量 970.2 亿立方米，农业用水量 3672.4 亿立方米，人工生态环境补水量 354.1 亿立方米。与上年比较，用水量减少 91.7 亿立方米，其中：生活用水量增加 4.1 亿立方米，工业用水量增加 1.8 亿立方米，农业用水量减少 108.9 亿立方米，人工生态环境补水量增加 11.3 亿立方米。

水资源节约。全国人均综合用水量为 419 立方米，农田灌溉水有效利用系数 0.576，万元国内生产总值（当年价）用水量 46.9 立方米，万元工业增加值（当年价）用水量 24.3 立方米。按可比价计算，万元国内生产总值用水量和万元工业增加值用水量分别比 2022 年下降 6.4%和 3.9%。全国非常规水源利用量 212.3 亿立方米，其中：再生水利用量 177.6 亿立方米，集蓄雨水利用量 10.8 亿立方米，淡化海水利用量 3.8 亿立方米，微咸水利用量 12.6 亿立方米，矿坑（井）水利用量 7.5 亿立方米。

5 防洪抗旱

2023年，洪涝灾害直接经济损失2445.75亿元（水利工程设施直接损失633.66亿元），占当年国内生产总值的0.19%。全国农作物受灾面积4633.29×10³公顷，绝收面积559.01×10³公顷，受灾人口5278.93万人，因灾死亡失踪309人，倒塌房屋13万间。河北、北京、黑龙江、河南、吉林、四川、天津等省（直辖市）受灾较重。全国农作物因旱受灾面积3803.70×10³公顷，绝收面积218.37×10³公顷，直接经济总损失205.51亿元❶。全国累计有275万城乡人口、278万头大牲畜因旱发生临时性饮水困难。全年完成抗旱浇地面积6099.71×10³公顷，抗旱挽回粮食损失64.15亿公斤，解决了272万城乡居民和257万头大牲畜因旱临时饮水困难。

---

❶ 2023年洪涝灾害直接经济损失、全国农作物受灾面积、绝收面积、受灾人口、因灾死亡失踪人口、倒塌房屋数量，农作物因旱受灾面积、绝收面积、直接经济总损失等数据来源于应急管理部。

全年中央下拨水利救灾资金 38.73 亿元，其中：用于防汛 35.28 亿元，用于抗旱 3.45 亿元。中央水利救灾资金在安全度汛隐患排查整治、防洪工程设施水毁修复、应急抗旱保供水、白蚁等害堤动物防治、水利工程震损修复等方面发挥了重要作用，为保障防洪安全、饮水安全、粮食安全提供了有力支撑。

6 水利改革与管理

河湖长制。全国 31 个省（自治区、直辖市）党委和政府主要负责同志担任省级总河长，省、市、县、乡、村 120 万名河湖长（含巡河员、护河员）上岗履职，省、市、县设立河长制办公室，专职人员超 1.8 万名。纵深推进"清四乱"常态化规范化，全国共清理整治"四乱"问题 1.73 万个，拆除违法违规建筑物 1998 万平方米，清理垃圾 2274 万吨，打击非法采砂船 2000 多艘。

水资源管理。2023 年批复 15 条跨省江河水量分配方案。截至 2023 年年底，组织制定了 171 条（个）跨省重要河湖、546 条（个）省内河湖生态流量保障目标，完成 168 个典型河湖、343 个已建水利水电工程生态流量核定。20 个省（自治区、直辖市）印发了地下水取水总量控制、水位控制"双控"指标。取用水管理专项整治行动全面完成，基本摸清全国 586 万个取水口的分布和取水情况，整改完成 427 万个取水口的违规取水问题。对黄河流域 13 个地表水超载地市 62 个地下水超采县暂停新增取水许可。全国已发放电子证照 63 万套。推进南水北调受水区地下水压采，累计压采地下水 79.18 亿立方米。对华北地区 7 个水系、40条（个）河湖实施生态补水，累计补水98.40亿立方米。

节约用水管理。2023年，制（修）订啤酒、氧化铝等5项工业服务业用水定额国家标准。加大计划用水管理力度，推动长江经济带年用水量1万立方米以上工业和服务业单位实现计划用水管理全覆盖。开展规划和建设项目节水评价6948个。公布第六批323个节水型社会达标县（区）名单。引导社会资本投入节水领域，推动实施合同节水管理项目204项。全年新发放"节水贷"804亿元。开展节水载体示范建设，公布省级节水型灌区194处，节水型工业企业4863家，节水型高校485所，联合公告74家重点用水企业、园区水效领跑者。

水资源调度。第一批55条跨省江河流域和7项已建调水工程全部开展调度，印发第二批30条开展水资源调度的跨省江河流域名录。下达西辽河、汉江流域年度水量调度计划，科学实施黄河、黑河、金沙江、淮河、西江、松花江干流、太湖等跨省江河流域和引黄入冀补淀、引滦等调水工程水资源调度，保障流域区域供水安全和生态安全。黄河实现连续24年不断流，黑河东居延海实现连续19年不干涸，永定河首次实现全年全线有水，西辽河干流水头近25年来首次到达通辽规划城区界，漳河179公里河道实现全线贯通，白洋淀水面面积保持在250平方公里以上。

运行管理。"十四五"以来，累计完成水库大坝安全鉴定37820

座，完成小型水库除险加固 11332 座，完成小型水库雨水情监测设施建设 37593 座，完成小型水库大坝安全监测设施建设 26113 座，实行小型水库专业化管护 48226 座，合理妥善实施水库降等报废 3722 座。2023 年，全国 1111 座大中型水库、1352 座大中型水闸、21626 公里 3 级以上堤防工程实现标准化管理，其中 110 处工程通过水利部评价。2023 年，完成大中型水闸安全鉴定 713 座，完成国有水库、堤防、水闸管理与保护范围划定 2003 座、17496 公里、2953 座。截至 2023 年年底，累计认定国家水利风景区 934 个，其中：水库型 396 个，河湖型 465 个，灌区型 34 个，水土保持型 39 个。

水价水权改革。2023年4月1日,《水利工程供水价格管理办法》《水利工程供水定价成本监审办法》正式实施。启动丹江口水利枢纽等水利工程供水定价成本监审和价格校核工作。截至2023年年底,累计实施农业水价综合改革面积约9亿亩以上。全国水权交易系统在7个流域管理机构、31个省(自治区、直辖市)和新疆生产建设兵团、5个计划单列市全部完成部署。中国水权交易所2023年完成水权交易5762单,交易水量5.39亿立方米,同比分别增长64%和116%。

水土保持管理。全国共审批生产建设项目水土保持方案9.60万个,涉及水土流失防治责任范围3.98万平方公里;5.90万个生产建设项目完成水土保持设施自主验收报备。常态化开展覆盖全国范围(港、澳、台数据暂缺)水土保持遥感监管,通过卫星遥感解译,组织现场复核,共认定并查处违法违规项目和活动1.33万个。开展国家水土保持示范创建,共评定123个示范;推动江西赣州、陕西延安、福建长汀、山西右玉、黑龙江拜泉等5个市(县)开展全国水土保持高质量发展先行区建设。

农村水利水电管理。截至2023年年底,25个省份累计创建绿色小水电示范电站1067座,长江经济带小水电清理整改全面完成,黄河流域及其他区域有序推进。全国4.1万余座小水电站基本按要求落实生态流量。积极推进农村水电站安全生产标准化建设,全国累计建成安全生产标准化电站5264座,其中一级电站128座、二级电站1892座、三级电站3244座。

水库移民管理。全国共完成大中型水利工程搬迁移民5.5万人。截至2023年年底，完成2022年度新增14.08万搬迁移民人口核定登记工作，涉及21省（自治区、直辖市）108座水库，大中型水库移民后期扶持人口为2546万人。陕西引汉济渭、西藏拉洛、四川亭子口等79项重大水利枢纽工程有序开展移民安置验收。水库移民人均可支配收入为2.1万元。

水利监督。出台推进水利工程建设安全生产责任保险工作的指导意见，组织全行业开展水利安全生产风险隐患等5项整治工作，全年排查整治事故隐患35.5万个；对风险较高、发生事故的地区和单位实施重点监管，水利行业累计管控危险源255.2万个。制定印发《水利部直属单位水利工程运行管理监督检查办法》，组织对部直管水库、水闸、堤防险工险段、淤地坝开展检查。全年有96家单位通过一级标准化达标创建评审。

依法行政。2023年4月1日，《中华人民共和国黄河保护法》正式实施。《中华人民共和国水法》《中华人民共和国防洪法》修改列入十四届全国人大常委会立法规划，《长江河道采砂管理条例》修改出台。2023年，水利部（包括部机关和各流域管理机构）共受理行政审批事项79638件，办结79381件。全国立案查处水事违法案件2.9万件，当年结案2.5万件，结案率86.2%；水利部共办结行政复议案件19件，办理行政应诉39件。

水利科技。2023年，立项实施国家重点研发计划"长江黄河等重点流域水资源与水环境综合治理""重大自然灾害防控与公共安全"等涉水重点专项项目14项、国家自然科学基金长江水科学研究联合基金项目32项。立项首批水利部重大科技计划项目243项。截至2023年年底，水利部共有国家和部级重点实验室33个（含筹建19个），国家和部级工程技术研究中心15个，国家和部级野外科学观测研究站7个。印发2023年度成熟适用水利科技推广清单，发布成果101项，立项水利技术示范项目35项。重新组建水利部科学技术委员会。发布水利技术标准16项，其中国家标准3项，行业标准13项。创新构建中国水利技术标准外文版体系，收录已翻译和急需翻译标准168项，组织完成10项水利标准英译本翻译审定。由水利部推荐的2项标准荣获2023年度标准科技创新奖，2人分别荣获标准大师奖和领军人才奖。

国际合作。2023年，积极参与联合国水大会，在北京举办第18届世界水资源大会。与新加坡、肯尼亚、荷兰等国开展政策对话与技术交流，与荷兰、新加坡续签合作谅解备忘录，举办水利多双边交流活动71场。参与第三届"一带一路"国际合作高峰论坛，5项重要成果纳入论坛务实合作项目清单。与越方续签关于相互交换汛期水文资料的谅解备忘录，纳入习近平总书记访问越南成果清单。完成对周边国家69个水文站的国际报汛工作。

7 水利行业状况

职工与工资。全国水利系统从业人员 65.8 万人，其中，全国水利系统在岗职工 62.6 万人。在岗职工中，部直属单位在岗职工 5.5 万人，地方水利系统在岗职工 57.1 万人。全国水利系统在岗职工工资总额为 773.4 亿元，年平均工资 12.3 万元。2013—2023 年职工与工资情况见表 2。

表 2 2013—2023 年职工与工资情况

项　目	2013年	2014年	2015年	2016年	2017年	2018年	2019年	2020年	2021年	2022年	2023年
在岗职工人数/万人	100.5	97.1	94.7	92.5	90.4	87.9	82.7	77.8	74.8	72.3	62.6
其中：部直属单位/万人	7.0	6.7	6.6	6.4	6.4	6.6	6.6	6.7	6.0	5.6	5.5
地方水利系统/万人	93.5	90.4	88.1	86.1	84.0	81.3	76.1	71.1	68.8	66.7	57.1
在岗职工工资/亿元	415.3	451.4	529.4	640.5	739.1	802.7	787.6	790.9	818.7	863.9	773.4
年平均工资/(元/人)	41453	46569	55870	69377	83534	91307	95236	102000	110000	120000	123479

表3 全国水利发展主要指标（2018—2023年）

指标名称	单位	2018年	2019年	2020年	2021年	2022年	2023年
1. 灌溉面积	10^3公顷	74542	75034	75687	78315	79036	80605
2. 耕地灌溉面积	10^3公顷	68272	68679	69161	69609	70359	71644
其中：本年新增	10^3公顷	828	780	870	1114	1228	1552
3. 万亩以上灌区	处	7881	7884	7713	7326		
其中：30万亩以上	处	461	460	454	450		450
4. 万亩以上灌区耕地灌溉面积	10^3公顷	33324	33501	33638	35499		
其中：30万亩以上	10^3公顷	17799	17994	17822	17868		
5. 农村自来水普及率	%	81	82	83	84	87	90
6. 除涝面积	10^3公顷	24262	24530	24586	24480	24129	25078
7. 水土流失治理面积	万平方公里	131.5	137.3	143.1	149.6	156.0	162.7
其中：本年新增	万平方公里	6.4	6.7	6.4	6.8	6.8	7.0
8. 水库	座	98822	98112	98566	97036	95296	94877
其中：大型水库	座	736	744	774	805	814	836
中型水库	座	3954	3978	4098	4174	4192	4230
9. 水库总库容	亿立方米	8953	8983	9306	9853	9887	9999
其中：大型水库	亿立方米	7117	7150	7410	7944	7979	8077
中型水库	亿立方米	1126	1127	1179	1197	1199	1210
10. 全年水利工程总供水量	亿立方米	6016	6021	5813	5920	5998	5907
11. 堤防长度	万公里	31.2	32.0	32.8	33.1	33.2	32.5
保护耕地	10^3公顷	41409	41903	42168	42192	41972	41748
堤防保护人口	万人	62837	67204	64591	65193	64284	63941
12. 水闸总计	座	104403	103575	103474	100321	96348	94460
其中：大型水闸	座	897	892	914	923	957	911
13. 年末全国水电装机容量	万千瓦	35226	35804	37028	39094	41350	42154
全年发电量	亿千瓦时	12329	13021	13553	13399	12020	

续表

指标名称	单位	2018年	2019年	2020年	2021年	2022年	2023年
14. 农村水电装机容量	万千瓦	8044	8144	8134	8290	8063	8157
全年发电量	亿千瓦时	2346	2533	2424	2241	2368	2303
15. 当年完成水利建设投资	亿元	6603	6712	8182	7576	10893	11996
按投资来源分：							
（1）中央政府投资	亿元	1753	1751	1787	1709	2218	2552
（2）地方政府投资	亿元	3260	3488	4848	4237	6004	6129
（3）国内贷款	亿元	753	636	614	699	1451	2023
（4）利用外资	亿元	5	6	11	8	6	
（5）企业和私人投资	亿元	565	588	690	718	1066	1292
（6）债券	亿元	42	10	87	104	75	
（7）其他投资	亿元	226	233	145	101	74	
按投资用途分：							
（1）防洪工程	亿元	2175	2290	2802	2497	3628	3946
（2）水资源工程	亿元	2550	2448	3077	2866	4474	5095
（3）水土保持及生态建设	亿元	741	913	1221	1124	1626	1446
（4）水电工程	亿元	121	107	92	79	107	79
（5）行业能力建设	亿元	47	63	85	80	124	131
（6）前期工作	亿元	132	133	157	137	730	244
（7）其他	亿元	836	757	747	794	2218	1055

说明：1. 本公报不包括香港特别行政区、澳门特别行政区及台湾省的数据。
　　　2. 农村水电的统计口径为单站装机容量为5万千瓦及以下的水电站。

2023 STATISTIC BULLETIN ON CHINA WATER ACTIVITIES

Ministry of Water Resources, People's Republic of China

The year of 2023 is the first year of fully implementing the essence of the 20th National Congress of the Communist Party of China. Over the past year, water resources departments at all levels, under the guidance of Xi Jinping' thoughts on socialism with Chinese characteristic in the new age, carried out the essence of the 20th National Congress of the Communist Party of China, and put the guidelines of General Secretary Xi Jinping of "prioritizing water conservation, advancing spatial equilibrium, taking systematic approaches and promoting government-market synergy", as well as important discourses on water governance into practices. We carried forward implementation of decisions and arrangements of the Central Committee of the Communist Party of China and the State Council, advanced new development concept in an accurate and comprehensive manner, coordinated high-quality development and high-level security, adhered to guidelines of water governance and bottom line thinking, took problem-oriented approaches, focused on prevention and systematic governance, boosted innovative development, achieved great victories in water and drought disaster prevention, and historic highs in investment and scale of water project construction, and made breakthrough progress in reform and innovation and smooth progress in all undertakings, and marked a solid step in promoting high-quality development of water resources in the new phase.

I. Investments in Water Resources Construction

In 2023, the total investment in water resources construction amounted to 1199.6 billion Yuan, among which, 908.9 billion Yuan was being allocated for construction projects, accounting for 75.8%; 58.3 billion Yuan for installation projects, accounting for 4.9%; 32.1 billion Yuan for expenditure on purchases of machinery, electric equipment and instruments, accounting for 2.7%; and 200.3 billion Yuan for other purposes, accounting for 16.6%. The completed investment in water resources construction during 2016–2023 is shown in Table 1.

Table 1 Completed investment in water resources construction during 2016–2023 unit: billion Yuan

	2016	2017	2018	2019	2020	2021	2022	2023
Total completed investment	610.0	713.2	660.3	671.2	818.2	757.6	1,089.3	1,199.6
Construction projects	442.2	507.0	487.7	498.8	601.5	585.1	849.2	908.9
Installation projects	25.5	26.6	28.1	24.3	32.0	33.0	48.6	58.3
Purchases of machinery, electric equipment and instruments	17.3	21.2	21.4	22.1	25.0	20.4	28.7	32.1
Others	125.0	158.4	123.1	126.0	159.7	119.1	162.8	200.3

According to the type of projects, 322.7 billion Yuan was allocated to the construction of river basin flood control projects, accounting for 27%; 566.5 billion Yuan was for key projects of national water network, accounting for 47%; 207.9 billion Yuan was for river and lake ecological restoration and improvement, accounting for 17%; and 102.5 billion Yuan for hydrological infrastructures and digital twin water resources construction, account for 9%. The completed investment of different types of projects in 2023 is shown in Figure 1.

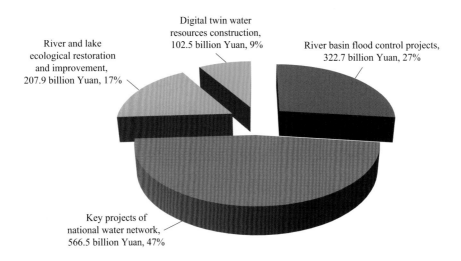

Figure 1　Completed investment of different types of projects in 2023

Of the total competed investment, the Central Government contributed 11.8 billion Yuan, and local governments contributed 1,187.8 billion Yuan. The investments completed by large and medium-sized projects reached 298.7 billion Yuan; and investment completed by small and other projects reached 900.9 billion Yuan. The investments for new projects were 916.7 Yuan, and investments completed by rehabilitation and expansion projects were 282.9 billion Yuan.

A total of 41,014 water projects were implemented in 2023, with a total investment of 5,479.7 billion Yuan. Among which, 27,925 new projects were launched in 2023, with an increase of 11.5%. The completed investment for water resources project construction during 2011–2023 is shown in Figure 2.

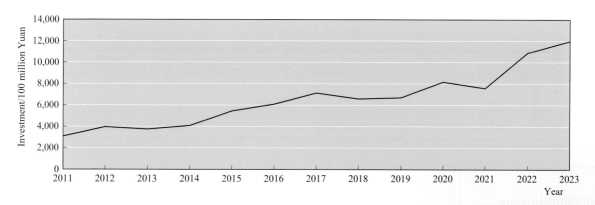

Figure 2　Completed investment for water resources project construction during 2011–2023

II. Construction of Key Water Resources Projects

River basin flood control projects. Yaojiaping and Fenghuangshan reservoirs in the Yangtze River Basin, Anqing river section harness of the Yangtze River, flood storage areas in Kangshan and Zhu Hu of Poyang Hu started construction. Expansion of Dongzhuang Reservoir in the Yellow River Basin and Wangyao Reservoir had been accelerated with full construction of downstream flood control works. Improvement of lowland in major plain of the Huai He Basin and flood storage areas along the Huai He began construction. Second phase of waterway for the estuary of the Huai He has been accelerated. In the Hai He Basin, harness of the Wenchaojian River in Beijing, North Canal of Tianjin, Hutuo River in Hebei and Fuyangxin River had been implemented. The construction of the Qingshan Reservoir was initiated. The construction of Wuzhou flood control embankment in the Pearl River Basin, Shiliudaogang Reservoir in the Songhua Jiang Basin, and follow-up works of western channel of the Tai Hu to expand drainage of Hangjia Hu were commenced. Improvement of Wusong Jiang was carried out in an orderly manner, and the whole line of the Tai Hu levees reached the standard.

Key projects of national water network. There were 44 major projects, including the backbone of Jilin water network, key project for increase of grain production capacity in Heilongjiang, the Xiong'an Canal, and water allocation around the Beibu Gulf of Guangxi, started construction. A group of key projects achieved key milestones, such as completion of main body of Xi Jiang Datengxia Multi-purpose Dam Project, trial operation of water diversion from the Han River to the Wei River and water diversion from the Yangtze River to the Huai He, whole-line water conveyance of the Guangdong Pearl River Delta water allocation scheme, acceleration of construction of the Yangtze River to supplement the Han Jiang and the Dianzhong Water Diversion Projects, and damming of the Huangzangsi Multi-purpose Dam Project for reservoir impoundment in the Hei He.

Rural water supply, irrigation and hydropower development. There were 5 large irrigation districts initiated, incluing Tengchong in Yunnan and Xialiujia in Guagnxi. Continuous and counterpart facility construction and rehabilitation for modernization in 598 large or medium-size irrigation districts had been accelerated, with a newly-increased or improved irrigated area of 70 million mu. The newly-constructed rural water supply projects totaled 23,000 that may safeguard water supply for 110 million rural residents. The percentage of rural population access to tap water rose to 90%. Large-scale water supply projects cover 60% of the rural population. There were about 4,000 hydropower stations equipped with intelligent system. Computer control systems or sub stations were installed in more than 200 small hydropower stations. There were 44 newly-increased hydropower stations with an installed capacity of 506,500 kW.

Key river and lake restoration and improvement. A water quality safeguarding system has been built for the Danjiangkou Reservoir Area and its upstream watershed. Ecological restoration of key rivers and lakes such as Mulanxi in Fujian, Chaganhu in Jilin, and Chaohu Lake in Anhui was accelerated. The Beijing-Hangzhou Grand Canal that had cut off for a century was reconnected. The Yongding River, which had been dried up for 26 years, achieved full-year water conveyance for the first time. Comprehensive management of groundwater over-exploitation in key areas such as North China had been intensified. The newly-increased areas for soil and water conservation and comprehensive control of soil erosion reached 70,000 km^2, of which the areas under the National Major Project for Soil Conservation amounted to 12,700 km^2. The slope farmland of 824,600 mu were harnessed. There were 600 newly-built silt retention dams and check dams. Up to 432 silt-retention dams with risk of hazard were strengthened and rehabilitated.

Digital twin water resources construction. The construction of digital twin watersheds, digital twin water networks, and digital twin projects has been boosted. The first key digital twin basin project in China a full-coverage water monitoring system of the Yangtze River Basin, was commenced. A first national level database with a data volume of billions of bytes (PB) was completed. Significant progress had been made in the construction of ministerial level water resources model platform, knowledge platform, and resource sharing platform. The "three defense lines" for rainfall and water condition monitoring and forecasting had been accelerated. The application of "prediction, early-warning, rehearsal and contingency planning" for river basin flood control realized a breakthrough. The optimization layout of national soil and water conservation monitoring stations approved for construction. Constant efforts had been made to advance pilots of digital twin irrigation areas.

III. Key Projects and Structures

Reservoirs and dams. The reservoirs built in China were 94,877, with a total storage capacity of 999.9 billion m^3. Of which large reservoirs were 836, with a total capacity of 807.7 billion m^3; medium-sized reservoirs were 4,230, with a total capacity of 121.0 billion m^3.

Embankments and water gates. The completed river dykes and embankments at Grade-V or above were 325,000 km. The accumulated length of dykes and embankments that met the standard reached 257,000 km, accounting for 79.0% of the total. Among which, up-to-the-standard dykes and embankments of Grade-I and Grade-II reached 40,000 km, taking up 88.3% of the total. The completed dykes and embankments nationwide could protect 640 million people and 41.748 million hm^2 of farmland. The completed water gates with a flow of 5 m^3/s were 94,460, of which 911 were large water gates. By type, there were 7,300 flood diversion sluices, 16,857 drainage/return water sluices, 4,522 tidal barrages, 12,461 water diversion intakes and 53,320 controlling gates.

Tube wells and pumping stations. There were 5.216 million tube wells with a daily water abstraction capacity equal to or larger than 20 m^3 or an inner diameter equal to or larger than 200 mm, completed for water supply in the whole country. A total of 91,324 pumping stations with a flow of 1 m^3/s or an installed voltage above 50 kW were put into operation, including 507 large, 5,014 medium-size and 85,803 small pumping stations.

Irrigation schemes. Irrigation districts with a designed area of 2,000 mu or above were 21,340 in total, covering 40.075 million hm^2 of irrigated farmland. Of which, 154 irrigation districts had an irrigated area of 500,000 mu or above, and 296 irrigation districts had an area from 300,000 to 500,000 mu. By the end of 2023, the total irrigated area amounted to 80.605 million hm^2. The irrigated area of cultivated land reached 71.644 million hm^2 that accounted for 56.1% of the total in China.

Rural hydropower and electrification. By the end of 2023, the number of hydropower stations built in rural areas was 41,114, with an installed capacity of 81.57 million kW, accounting for 19.4% of the national total. The annual power generation by these hydropower stations reached 230.3 billion kWh.

Soil and water conservation. The area with soil conservation measures reached 1.627 million km^2; and the forbidden area for ecological restoration accumulated to 325,000 km^2. In 2023, dynamic monitoring had been continued for soil and water loss in all administrative areas above county level, key areas, and major river basins in the country, in order to collect comprehensive data on dynamic changes.

Hydrological station networks. There were 127,035 hydrological stations of all kinds constructed in the whole country, including 3,312 national basic hydrologic stations, 5,169 special hydrologic stations, 20,633 gauging stations, 56,279 precipitation stations, 9 evaporation stations, 26,576 groundwater monitoring stations, 9,187 water quality stations, 5,809 soil moisture monitoring stations and 61 experimental stations. Among them, 84,921 stations of various kinds can provide hydrological information to water administration authorities at and above county level; 2,575 stations can release forecast and 2,729 can release early warnings; 3,537 equipped with online flow measurement and 7,379 equipped with video monitoring system. A water quality monitoring system consisted of 330 monitoring centers, sub-centers and water quality stations (sections) at central, basin, provincial and local levels was set up.

Water networks and information systems. By the end of 2023, water resources departments and authorities at and above provincial level were equipped with 10,495 servers of varied kinds, forming a total storage capacity of 58.32 PB, and keeping 6.37 PB of data and information. Water resources departments and authorities at and above county level had equipped with 3,766 sets of various kinds of satellite equipment, 10,809 flood forecasting stations for short message transmission from the Beidou Satellites, 49 vehicles for emergency communication, 2,827 cluster communication terminals, 464 narrowband and broadband communication systems, and 3,351 unmanned aerial vehicles (UAV). A total of

603,500 information gathering points were available for water resources departments and authorities at and above county level, including 303,500 points for collecting data of hydrology, water resources and soil and water conservation and 300,000 points for safety monitoring at large and medium-sized reservoirs.

IV. Water Resources Conservation, Utilization and Protection

Water resources conditions. In 2023, the total water resources in the country amounted to 2,578.25 billion m^3, 6.6% less than the normal years. The mean annual precipitation was 642.8 mm, 0.2% less than the normal years and 1.8% more than the previous year. The total storage of 768 large and 3,942 medium-sized reservoirs were 459.45 billion m^3, increasing 39.04 billion m^3 comparing to that at the beginning of the year.

Water resources development. In 2023, the newly-increased capacity of water supply by water structures above designated size❶ was 9.18 billion m^3. By the end of 2023, the total capacity of water supply reached 906.2 billion m^3, among which 67.78 billion m^3 was from water supply systems at the county level, 247.41 billion m^3 from reservoirs, 211.97 billion m^3 from water diversion of rivers and lakes, 185.26 billion m^3 from pumping stations of rivers and lakes, 137.26 billion m^3 from electro-mechanical wells, 37.19 billion m^3 from ponds, weirs and cellars, and 19.33 billion m^3 from unconventional water sources.

Water resources utilization. In 2023, the total quantity of water supply amounted to 590.65 billion m^3, including 487.47 billion m^3 from surface water, 81.95 billion m^3 from groundwater and 21.23 billion m^3 from other sources. The total water consumption amounted to 590.65 billion m^3, including 90.98 billion m^3 of domestic water use, 97.02 billion m^3 of industrial water use, 367.24 billion m^3 of agricultural

---

❶ Water projects above designated size include reservoirs with a total capacity of 100,000 m^3 or higher, pump stations with an installed flow at or above 1 m^3/s or an installed capacity at or above 50 kW, water diversion gates with a flow at or above 1 m^3/s, electric irrigation wells 200 mm or larger in inner diameter, and electro-mechanical wells with a water supply capacity at or above 20 m^3/d.

water use and 35.41 billion m³ for environment and ecological use. Comparing to the previous year, the total water consumption reduced by 9.17 billion m³, among which domestic water use increased 410 million m³ and industrial water use increased 180 million m³, agricultural water use decreased 10.89 billion m³ and artificial recharge for environmental and ecological use increased 1.13 billion m³.

Water resources conservation. The mean annual water consumption was 419 m³ in China. The coefficient of effective use of irrigated water was 0.576. Water use per 10,000 Yuan of GDP (at comparable price of the same year) was 46.9 m³ and that per 10,000 Yuan of industrial value added (at comparable price of the same year) was 24.3 m³. Based on the estimation at comparable prices, water uses per 10,000 Yuan of GDP and per 10,000 Yuan of industrial value added reduced by 6.4% and 3.9% respectively over the previous year. The consumption of unconventional water sources amounted to 21.23 billion m³, among which 17.76 billion m³ from reclaimed water, 1.08 billion m³ from collected or stored rainwater, 380 million m³ from desalinated water, 1.26 billion m³ from blackish water and 750 million m³ from treated mine water.

V. Flood Control and Drought Relief

In 2023, the direct economic loss caused by flood and waterlogging disasters was 244.575 billion Yuan (including 63.366 billion Yuan of direct losses of water facilities), accounting for 0.19% of the GDP in the same year. A total of 4,633,290 hm² of cultivated land were affected by floods, including 559,010 hm² no harvest farmland, affected population of 52.7893 million, 309 dead and missing and 130,000 collapsed houses. Provinces or municipalities of Hebei, Beijing, Heilongjiang, Henan, Jilin, Sichuan and Tianjing suffered heavily from severe flooding. The affected areas of farmland by drought were 3,803,700 hm², and the areas with no harvest were 218,370 hm², with a total of 20.551 billion Yuan of direct economic loss[1]. A total of 2.75 million urban and rural residents and 2.78 million man-feed big animals and livestock suffered from temporary drinking water shortage. Up to 6,099,710 hm² of lands were irrigated against droughts that retrieved a grain loss of 6.415 billion kg. Drinking water was provided to 2.72 million rural and urban population and 2.57 million big animals and livestock in order to alleviate temporary water shortage.

In, 2023, the Central Government allocated 3.873 billion Yuan for the mitigation of water-related disasters, including 3.528 billion Yuan for flood defense and 0.345 billion Yuan for drought relief. The funds for disaster relief from the Central Government have played a key role in ensuring safety during the flood season, hazard investigation and mitigation, repairing of damaged structures and facilities and emergency water supply, and provide strong support for safeguarding flood

---

[1] The data of direct economic losses caused by floods, the affected area of crops, the area of no harvest, the affected population, the number of dead and missing due to disasters, and the number of collapsed houses in 2023 all come from the Ministry of Emergency Management.

control and drinking water security and grain security.

VI. Water Reform and Management

River (lake) chief system. The governors in 31 provinces, autonomous regions and municipalities served as the general chiefs of rivers and lakes. About 1.2 million river (lake) chiefs (including river patrol officers or river rangers) at provincial, city, county and township levels stated to perform their duties. The offices of river chief system were set up at provincial, municipal and county levels, with more than 18,000 full-time employees. All-round efforts had been made in normalization and standardization of actions for rectifying "misappropriation, illegal sand excavation, disposing of wastes and building structures without permission", with a total of 17,300 illegal activities corrected, 19.98 million km^2 of illegally-constructed buildings removed, 22.74 million tons of garbage in river courses cleared up, and more than 2000 illegal sand-mining ships were punished.

Water resources management. In 2023, a total of 15 cross-province water allocation plans were approved. Targets for securing ecological flow of 171 cross-province major rivers and lakes were specified. Define the targets of securing ecological flow of 546 major rivers and lakes. Approval of ecological flows for 168 typical rivers and lakes and 343 existing water resources and hydropower projects was completed. The total quantity of groundwater withdrawal and indicators of "double control" on total groundwater use were issued by 20 provinces (autonomous regions or municipalities). A special-subject rectification action on water abstraction management has been completed, with a basic understanding on distribution and water abstraction of 5.86 million water intakes nationwide and rectification of illegal incidents of 4.27 million water intakes. The issue of new water abstraction license was terminated in 13 cities and 62 counties in the Yellow River Basin with groundwater over-abstraction. In 2023, the issued electronic licenses reached 630,000 in the whole country. Reduction of groundwater withdrawal in water receiving areas of South-to-North Water Diversion Project had been implemented, with an accumulated reduction of 7.918 billion m^3 of groundwater withdrawal. Ecological water replenishment was implemented for 7 water systems and 40 rivers and lakes in North China, with a cumulative replenishment of 9.84 billion cubic meters of water.

Water saving and water use management. In 2023, national standards were issued for water use quota in 5 industry and service sectors including bear and alumina. More emphasis had placed on planed water use that covered all enterprises of industrial and service sectors in the Yangtze River Basin, that have an annual water consumption of 10,000 m^3 and above. Water-saving evaluation was conducted for 6,948 plans and construction projects. The name-list of up-standard water-saving society at the county level of 323 counties or districts in six batches was announced. Introduction of social investment and input to water-saving field has been encouraged to promote the implementation of 204 contract-based water conservation and management projects. The newly-issued "water-saving loan" in 2023 reached 80.4 billion Yuan. Water-saving carrier pilots were implemented, resulting in 194 water-saving irrigation districts, 4,863 water-saving industrial enterprises, 485 water-saving universities and colleges. Major water users of 74 enterprisers and industrial parks were nominated as "water efficiency leader".

Water resources dispatching and allocation. In 2023, the first group of 55 cross-province river basins and 7 completed water diversion projects started water resources dispatching and allocation. The name-list of second group of 30 cross-province river basins was issued. The annual plans of water resources dispatching and allocation of the Xiliao He and Han Jiang were approved. Water resources dispatching for water diversion from the Yellow River to Baiyang Dian Wetlands and from the Luan River, and projects in cross-province river basins of Yellow River,

Hei He, Jinsha Jiang, Huai He, Xi Jiang, Mainstream of the Songhua Jiang and Tai Hu were carried out in scientific manner, in order to safeguard water supply safety and ecological safety in the regions of river basins. The Yellow River has kept flowing for 24 years and Dongjuyanhai of Hei He has kept flowing for 19 consecutive years. The Yongding He realized full-line water conveyance for the first time. The water head of Xiliao He mainstream reached to the planned section of Tongliao for the first time in 25 years. The whole-line water conveyance was realized in 179 km river courses of the Zhang He. The water surface area of the Baiyang Dian wetlands had kept above 250 km^2.

Operation and management. Since the implementation of the 14th Five-Year Plan, safety appraisal had been undertaken for 37,820 reservoirs. A total of 11,332 small reservoirs had been reinforced, with 37,593 rainwater monitoring facilities installed; 26,113 dam safety monitoring facilities installed; 48,226 professionally managed and protected; and 3,722 reservoirs downgraded and removed. In 2023, standardized management had been adopted in 1,111 large and medium-size reservoirs, 1,352 large and medium size water gates and 21,626 km of dykes and embankments above level 3 across the country, among which 110 projects passed evaluation of the Ministry of Water Resources. Safety evaluation was completed for 713 large and medium-sized water gates. Delimitation of administrative boundaries had been completed for 2,003 reservoirs, 17,496 kilometers of embankments, and 2,953 water gates. As of the end of 2023, a total of 934 national water scenic spots of various kinds had been approved, including 396 reservoir type, 465 river and lake type, 34 irrigation districts, and 39 soil and water conservation areas.

Water pricing reform. In 1 April 2023, *the Management Measures for Price of Water Supply of Water Resources Projects* and *the Supervision and Examination Measures for Pricing and Costs of Water Supply of Water Resources Projects* came into force. The work of supervision and appraisal of pricing and costs of water supply was initiated for water resources projects of Danjiangkou and so on. As of the end of 2023, reform of water pricing system had been applied to 0.9 billion mu of agricultural land accumulatively. Water right trading system has been applied in 7 river basin commissions, 31 provinces, autonomous regions and municipalities and Xinjiang Production and construction Corps and 5 specifically designated cities. China Water Exchange had completed a total 5,762 entitlement trading, with an amount of 539 million m^3 of water, with an increase of 64% and 116% respectively.

Soil and water conservation. In 2023, MWR approved 96,000 soil and water conservation plans of construction projects, covering an area of 39,800 km^2 within the scope of responsible for erosion control. Soil conservation facilities of 59,000 construction projects passed self-check and acceptance. Remote sensing was adopted in the whole country (except Hong Kong, Macao and Taiwan) to monitor soil and water loss caused by human activities in production and construction. On-site reinvestigations were organized according to the interpretation of data from remote sensing satellites, and 13,300 noncompliant projects were identified and punished. National demonstration areas for soil and water conservation were created, and 123 pilots were evaluated and selected. High-quality development has been promoted in the pilot cities or counties, namely Ganzhou of Jiangxi Province, Yan'an of Shaanxi Province, Changting of Fujian Province, Youyu of Shanxi Province and Baiquan of Heilongjiang Province.

Rural hydropower management. By the end of 2023, 1,067 small demonstrate hydropower stations in 25 provinces met the criteria of green small hydropower station. Removal and rehabilitation of small hydropower stations in the Yangtze Economic Belt was completed. Orderly promotion was made in the Yellow River Basin and other regions. About 41,000 small hydropower stations had function of ecological flow release. Standards for safe production and operation had been applied to hydropower stations in rural areas, with 5,264 hydropower stations complied with relevant standards, including 128 of level A, 1,892 of level B and 3,244 of level C.

Resettlement of reservoir projects. The relocated population of large and medium-size water resources projects reached 55,000. By the end of 2023, approved registration of newly-added 140,800 population in 2022 was completed that related to 108 reservoirs in 21 provinces (autonomous region or municipality). The population of large and medium-size water resources projects with follow-up support reached 25.46 million. Check and acceptance of resettlement arrangements for 79 key multi-purpose dam projects, including water diversion from Han Jiang to Wei He, Laluo in Xizang and Tingzikou in Sichuan had been conducted in an orderly manner. The per capita disposable income of reservoir resettlement reached 21,000 Yuan.

Supervision. The guidelines of introducing liability insurance for water resources project construction and production safety were issued. Actions had been taken to investigate 5 types of hidden dangers and risks associated with production safety, and 355,000 accident potentials were identified and rectified. The supervision had focused on the area or unit that had high risks and accidents occurred before. There were accumulative 2.552 million potential hazards in the water sector were brought under control. *Inspection Measures for Operation and Management of Water Resources Project within the Jurisdiction of Units Directly under the Ministry of Water Resources* was formulated and issued. Inspections were organized and conducted for reservoirs, water gates, hazard embankments and sections and structures within the jurisdiction of Units Directly under the Ministry of Water Resources. There were 96 units passed the first level up-to-standard inspections in the whole country.

Law-based administrative and enforcement. In 1 April 2023, *Yellow River Protection Law of the People's Republic of China* was passed and adopted. The 14th National People's Congress Standing Committee included the revision of *Water Law of the People's Republic of China* and *Flood Control Law of the People's Republic of China* in the legislative plan. *Regulations for Yangtze River Sand Mining Management* was revised and promulgated. MWR (including river basin authorities) handled 79,638 water-related cases and resolved 79,381 cases. There were 29,000 water-related cases were put in file and invested in the whole country and 25,000 resolved with a rate of 86.2%. MWR handled and concluded 19 administrative reconsideration cases and 39 administrative proceedings.

Water science and technology. In 2023, a total 14 water-related major special subject projects were approved and implemented, including Comprehensive Regulation of Water Resources and Water Environment in Major River Basins of the Yangtze River and the Yellow River and Prevention and Control of Major Natural Disasters and Public Security in the list of national key research and development plan. There were 32 projects with joint funds for scientific studies on the Yangtze River and the Yellow River were approved and implemented. There were 243 projects approved as the first group of major water science and technologies projects. By the end of 2023, MWR had 33 (including 19 in preparing) national level or ministerial level labs, 15 national and ministerial level engineering technology research centers and 7 national and ministerial level field scientific observation and research stations. The name-list of applicable water technologies for scientific extension and 101 achievements in 2023 were released. There were 35 water science and technologies demonstration projects approved. Science and Technology Committee of the Ministry of Water Resource was reorganized. There were 16 water-related technical standards made public, among 3 national standard, 13 industry standard. Foreign language edition of water resources standards of China was created innovatively, with 168 standards included in the list of urgently needed

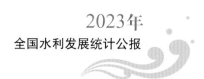

for translations. There were 10 English version of water resources standards completed approval. Among the nominated standards of MWR, 2 standards received 2023 standard science and technology innovation award. 2 peoples received Master Award and Leading Talent Award.

International cooperation. In 2023, MWR participated in the UN 2023 Water Conference and organized the 18th World Water Congress. Policy dialogue and technical exchanges were held with counterparts from Singapore, Kenya, the Netherlands and other countries. MOUs for continuous cooperation were signed with the Netherlands and Singapore. MWR organized 71 multi-bilateral exchange activities. At the Third Belt and Road International Cooperation Summit Forum, there were 5 main achievements of MWR included in the list of Practical Achievements of the forum. MWR renewed the MOU between China and Vietnam on exchanging hydrological data during flood season, included in the list of achievements during the visit of general secretary Xi Jinping to Vietnam. Flood warning was reported and delivered to 69 hydrological stations in the neighboring countries.

VII. Current Status of the Water Sector

Employees and salaries. Employees of the water sector totaled 658,000. Of which, in-service staff members amounted to 626,000, including 55,000 working in agencies directly under the Ministry of Water Resources, and 571,000 working in local agencies. The total salary of in-service staff members nationwide was 77.34 billion Yuan, and the annual average salary per person was 123,479 Yuan. Employees and salaries during 2013 – 2023 are shown in Table 2.

Table 2 Employees and salaries during 2013 – 2023

Composition of Funds	2013	2014	2015	2016	2017	2018	2019	2020	2021	2022	2023
Number of in service staff /10^4 persons	100.5	97.1	94.7	92.5	90.4	87.9	82.7	77.8	74.8	72.3	62.6
Of them: staff of MWR and agencies under MWR /10^4 persons	7.0	6.7	6.6	6.4	6.4	6.6	6.6	6.7	6.0	5.6	5.5
Local agencies /10^4 persons	93.5	90.4	88.1	86.1	84.0	81.3	76.1	71.1	68.8	66.7	57.1
Salary of in-service staff /10^8 Yuan	415.3	451.4	529.4	640.5	739.1	802.7	787.6	790.9	818.7	863.9	773.4
Average salary /(Yuan/person)	41,453	46,569	55,870	69,377	83,534	91,307	95,236	102,000	110,000	120,000	123,479

Table 3 Main Indicators of National Water Resources Development (2018-2023)

Indicators	Unit	2018	2019	2020	2021	2022	2023
1. Irrigated area	10^3 hm^2	74,542	75,034	75,687	78,315	79,036	80,605
2. Farmland irrigated area	10^3 hm^2	68,272	68,679	69,161	69,609	70,359	71,644
Among which: Newly-increased of the year	10^3 hm^2	828	780	870	1,114	1,228	1,552
3. Irrigation districts with an irrigated area above 10,000 mu	unit	7,881	7,884	7,713	7,326		
Among which: Irrigation districts with an irrigated area above 300,000 mu	unit	461	460	454	450		450
4. Farmland irrigated areas in irrigation districts over 10,000 mu	10^3 hm^2	33,324	33,501	33,638	35,499		
Among which: Farmland irrigated areas in irrigation districts over 300,000 mu	10^3 hm^2	17,799	17,994	17,822	17,868		
5. Rural population accessible to safe drinking water	%	81	82	83	84	87	90
6. Waterlogging control areas	10^3 hm^2	24,262	24,530	24,586	24,480	24,129	25,078
7. Soil erosion control area	10^4 km	131.5	137.3	143.1	149.6	156.0	162.7
Among which: newly-increased areas	10^4 km	6.4	6.7	6.4	6.8	6.8	7.0
8. Reservoirs	unit	98,822	98,112	98,566	97,036	95,296	94,877
Among which: Large-sized	10^4 km^2	736	744	774	805	814	836
Medium-sized	10^4 km^2	3,954	3,978	4,098	4,174	4,192	4,230
9. Total storage capacity	10^8 m^3	8,953	8,983	9,306	9,853	9,887	9,999
Among which: Large-sized	10^8 m^3	7,117	7,150	7,410	7,944	7,979	8,077
Medium-sized	10^8 m^3	1,126	1,127	1,179	1,197	1,199	1,210
10. Total water supply capacity of water projects in a year	10^8 m^3	6,016	6,021	5,813	5,920	5,998	5,907

Continued

Indicators	Unit	2018	2019	2020	2021	2022	2023
11. Length of dikes and embankments	10^4 km	31.2	32.0	32.8	33.1	33.2	32.5
Cultivated land under protection	10^3 hm^2	41,409	41,903	42,168	42,192	41,972	41,748
Population under protection	10^5 peoples	62,837	67,204	64,591	65,193	64,284	63,941
12. Total water gates	unit	104,403	103,575	103,474	100,321	96,348	94,460
Among which: large gates	unit	897	892	914	923	957	911
13. Total installed capacity by the end of the year	10^4 kW	35,226	35,804	37,028	39,094	41,350	42,154
Yearly power generation	10^8 kWh	12,329	13,021	13,553	13,399	12,020	
14. Total installed capacity of rural hydropower	10^4 kW	8,044	8,144	8,134	8,290	8,063	8,157
Yearly power generation	10^8 kWh	2,346	2,533	2,424	2,241	2,368	2,303
15. Completed investment of water projects	10^8 Yuan	6,603	6,712	8,182	7,576	10,893	11,996
Divided by different sources							
(1) Central government investment	10^8 Yuan	1,753	1,751	1,787	1,709	2,218	2,552
(2) Local government investment	10^8 Yuan	3,260	3,488	4,848	4,237	6,004	6,129
(3) Domestic loan	10^8 Yuan	753	636	614	699	1,451	2,023
(4) Foreign funds	10^8 Yuan	5	6	11	8	6	
(5) Enterprises and private investment	10^8 Yuan	565	588	690	718	1,066	1,292
(6) Bonds	10^8 Yuan	42	10	87	104	75	
(7) Other sources	10^8 Yuan	226	233	145	101	74	

Continued

Indicators	Unit	2018	2019	2020	2021	2022	2023
Divided by different purposes:							
(1) Flood control	10^8 Yuan	2,175	2,290	2,802	2,497	3,628	3,946
(2) Water resources	10^8 Yuan	2,550	2,448	3,077	2,866	4,474	5,095
(3) Soil and water conservation and ecological recovery	10^8 Yuan	741	913	1,221	1,124	1,626	1,446
(4) Hydropower	10^8 Yuan	121	107	92	79	107	79
(5) Capacity building	10^8 Yuan	47	63	85	80	124	131
(6) Early-stage work	10^8 Yuan	132	133	157	137	730	244
(7) Others	10^8 Yuan	836	757	747	794	2,218	1,055

Notes:

1. This Bulletin do not include the data of Hong Kong, Macao and Taiwan.

2. The statistics of rural hydropower refers to the hydropower stations with a per unit installed capacity of 50,000 kW or less.

《2023年全国水利发展统计公报》编辑委员会

主　　　　任：陈　敏
副　主　　任：吴文庆　张祥伟
委　　　　员：（按姓氏笔画排序）
　　　　　　　邢援越　巩劲标　朱东恺　任骁军　刘宝军　李　烽
　　　　　　　李兴学　李春明　李原园　杨卫忠　吴　强　张　彬
　　　　　　　张玉山　张新玉　陈茂山　郑红星　赵　卫　姜成山
　　　　　　　袁其田　夏海霞　钱　峰　倪　莉　郭孟卓　曹纪文
　　　　　　　彭　静　谢义彬

《2023年全国水利发展统计公报》主要编辑人员

主　　　　编：张祥伟
副　主　　编：谢义彬　吴　强　汪习文
执　行　编　辑：张光锦　徐　吉　李　淼　张　岚
主要参编人员：（按姓氏笔画排序）
　　　　　　　万玉倩　马　超　王小娜　王鹏悦　田　枞　刘　阳
　　　　　　　刘　品　孙宇飞　杜崇玲　李笑一　杨　波　吴泽斌
　　　　　　　吴海兵　吴梦莹　邱立军　张贤瑜　张晓兰　欧阳珊
　　　　　　　周哲宇　房　蒙　聂少安　殷　殷　殷海波　高立军
　　　　　　　郭　悦　黄藏青　蒋雨彤　韩绪博　谢雨轩　潘利业
英　文　翻　译：谷丽雅

◎ 主编单位
水利部规划计划司

◎ 协编单位
水利部发展研究中心

◎ 参编单位
水利部办公厅
水利部政策法规司
水利部财务司
水利部人事司
水利部水资源管理司
全国节约用水办公室
水利部水利工程建设司
水利部运行管理司
水利部河湖管理司
水利部水土保持司
水利部农村水利水电司
水利部水库移民司
水利部监督司
水利部水旱灾害防御司

水利部水文司
水利部三峡工程管理司
水利部南水北调工程管理司
水利部调水管理司
水利部国际合作与科技司
水利部综合事业局
水利部信息中心
水利部水利水电规划设计总院
中国水利水电科学研究院